# Barns of Will County

AF772055

A collection of photographs by

Kevin McNulty, Sr.

Copyright © 2021 KMC Publishing Company - 2nd Edition
All Rights Reserved
All photography property of Kevin McNulty, Sr
ISBN 978-1-7324091-1-8

# Will County Illinois

Will County was formed in 1836, extending from current Cook County through Iroquois County. In addition to its present area, the County included the are of current-day Kankakee County, Illinois north of the Kankakee River. Will County lost that area when Kankakee County was organized in 1852. Since then its boundaries have been unchanged.

The story of rural Will County is one of an encroaching suburban sprawl and the fading presence of the family farm. It is not an uncommon thing to see in the Midwest, particularly in areas at the fringe of a major American city like Chicago.

While most of the northern areas of the county such as Bolingbrook and the southern reaches of Naperville are congested suburban towns, the area around the county seat of Joliet has its roots in the merger of the Des Plaines and Kankakee Rivers which pour into the Illinois which arrives at the Mississippi at the Quad Cities.

The history of the Joliet area is one of commerce and industry, going back to when rivers were the sole means of transportation. Today, commerce continues at the mouth of the Illinois River where the two feeders merge. But a new modern merging of rail lines and highways make Will County a staging area for "closer in" buildings of distribution that feed the metropolitan area and spread east and west to other Midwestern states.

Joliet was a planned suburban industrial center, part of a crescent-shaped ring around old Chicago when manufacturing was king, and the smoke of manufacturing was placed some distance from the central city. The old arc of industry began with Waukegan in the north, then extended around Chicago's center at Elgin, Aurora, Joliet, and finally, Chicago Heights, which is directly east of Joliet, fifteen miles short of the Indiana border. And so, like its industrial "sister cities", Joliet on the western end of Will County leaves a mark at its perimeter that is not rural in nature.

The western boundary, just past Joliet, reaches north until it ends short of the Aurora influence. It also extends south of Joliet through the unique areas of preserve wherein the Midewin National Tallgrass Prairie lies and buffalo occupy the newly protected area. In the far southwest corner, lies the Braidwood Nuclear Generating Station, capping off the wildly diverse nature of the western edge of Will.

The old Route 6 and Lincoln Highway extend east from Joliet to the Indiana border and as it does, the border of Cook and Will county gradually merge along these roads beginning at Tinley Park in the form of Interstate 80, where modern travelers now pass.

It is south of this line where one can still see vestiges of the old rural days. Close to the suburban-rural border, we have described, one sees fields, dotted with non-farm housing where city dwellers and now ex-urban refugees have arrived - escaping "life in the city" - while remaining productive on the broadband highways of today's commerce.

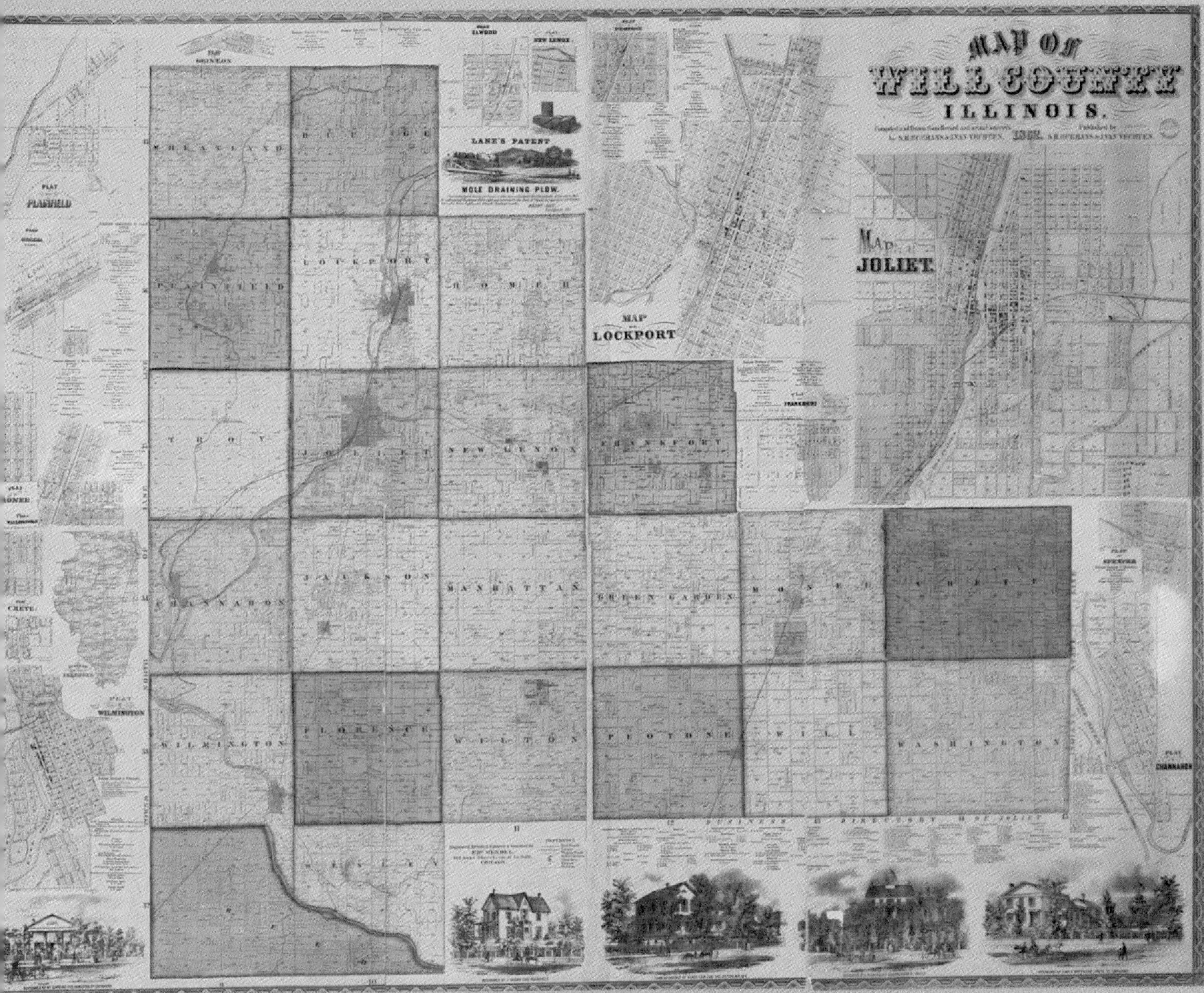


**Map of Will County, Illinois**

Library of Congress

S. H. Burhans and J. Van Vechten, 1862.

# A Note from the Author

On March 6, 2020, I left a substitute teaching position near my home, where I had just spent three months subbing in the orchestra department. I had the opportunity to work with three orchestras and our concert was scheduled for March 9. The concert never happened. School closed on the 8th. There was talk in the media about a virus that was showing up in cities around the U.S. Nonetheless, my term had ended, so I returned to my home-based business once again. I didn't know it then, but this new virus would change a whole lot of things. It would also give me time again to write. I have always written music since college but my increased exposure to other's good music has kept me away from the piano more days than not. Somewhere in the late '80s, I began to write more words than notes. The process is similar, just different elements. Writing music is a bit more gut-wrenching.

Around 2014, my interest in family history brought me to the farm of my relatives, the Bouchards in Kankakee County, Illinois. Two sisters and the brother of my grandmother were the last to live there but the fields were taken over by cousins toward the end. I decided to get a photo of their old barn before it was either razed or fell on its own. Not being a professional photographer like my brother Chuck, I knew little about photography. But I noticed how easy it was to take pictures of old barns. They have deep character and tell a story. You cannot see an old barn without picturing hard work and how rural life must have been in America a long time ago.

So, with little photographic knowledge, I began taking pictures of other barns, too. The more photos I took, the more I enjoyed it. I enjoyed the conversations with the people living on the old properties, as well. Many just lived in the house allowing others to farm the land. But on occasion, some fourth and fifth generations could be found on the same farm, in the same house. I took pictures of more farms. At each chance, I drove a different country road - shooting barns, sheds, animals, and many other trappings still present on farms today. When I was done, I had traveled up and down every road in the county. Other than the river that cuts a natural path through the county, roads are organized in a grid fashion. It was easy to map one's progress, and it was remarkable. No radio. No Music, No phone. Just open, flat Kankakee County. I published "Barns of Kankakee County" in 2014.

And so, we return to 2020 and COVID-19. Like many, I spent an entire year at home. The necessity of "going places" did not matter much to me. With nowhere to go, I took the opportunity to drive Will County and take pictures of barns for a whole year. It was a vastly different landscape and quite a different experience from Kankakee. I hope you can sense the beauty and wonder of what I saw with my eyes and attempted to capture.

While I now live in the congested area of Chicago, I am always drawn to the two counties to my south. Fortunately, I need only drive two miles in that direction, and I am in the country. I want to dedicate this year-long effort to the people at another place I have not been for a year. They are good Will County folks. Thanks to Brian, Kat, Rita, Chris, Donna, Sissy, Ruben, and yes, Phil, the crew, and all my friends at Brian's Place in Monee. Thanks for being pros and providing me a place to go to "get out of the city." Readers should drive the country roads of Will County, someday. When you're done, stop in at "Brian's." You will realize you're on familiar ground.

# Barns of Will County

In the interest of maintaining the integrity of each photo in this book, the author has chosen not to number each page. The book is organized by seasons beginning with Spring and ending in Winter. There are also special themed sections in the book. The reader can experience and enjoy the photos herein without regard to sequence.

Hundreds of years ago, many farmers would seal their barns with linseed oil, which was an orange-colored oil derived from the seeds of the flax plant. To this oil, they would add a variety of things including rust. Rust was plentiful on farms, and it killed fungi and mosses. It also turned the mixture red in color.

# Remaining Reds

# Little Red Garage

1884

In the beginning, farmers in the East preferred to use flaxseed oil and white lead to paint in their barns. Lead was added to accelerate drying, increase durability, and resist moisture. Early on, farmers in the Midwest had less access to lead so they mixed red rust with their flaxseed oil to kill fungi, resist moisture and lessen wood decay. The red barn tradition never took root along the path of the National Road, however, because freight, including white lead, was accessible to nearby farmers. One found more of a white barn tradition in Southern Illinois due to this. But while Will County has a majority of red barns, white barns persist also partially due to the county's proximity to Chicago - the rail center of America after the Civil War.

# White Barns

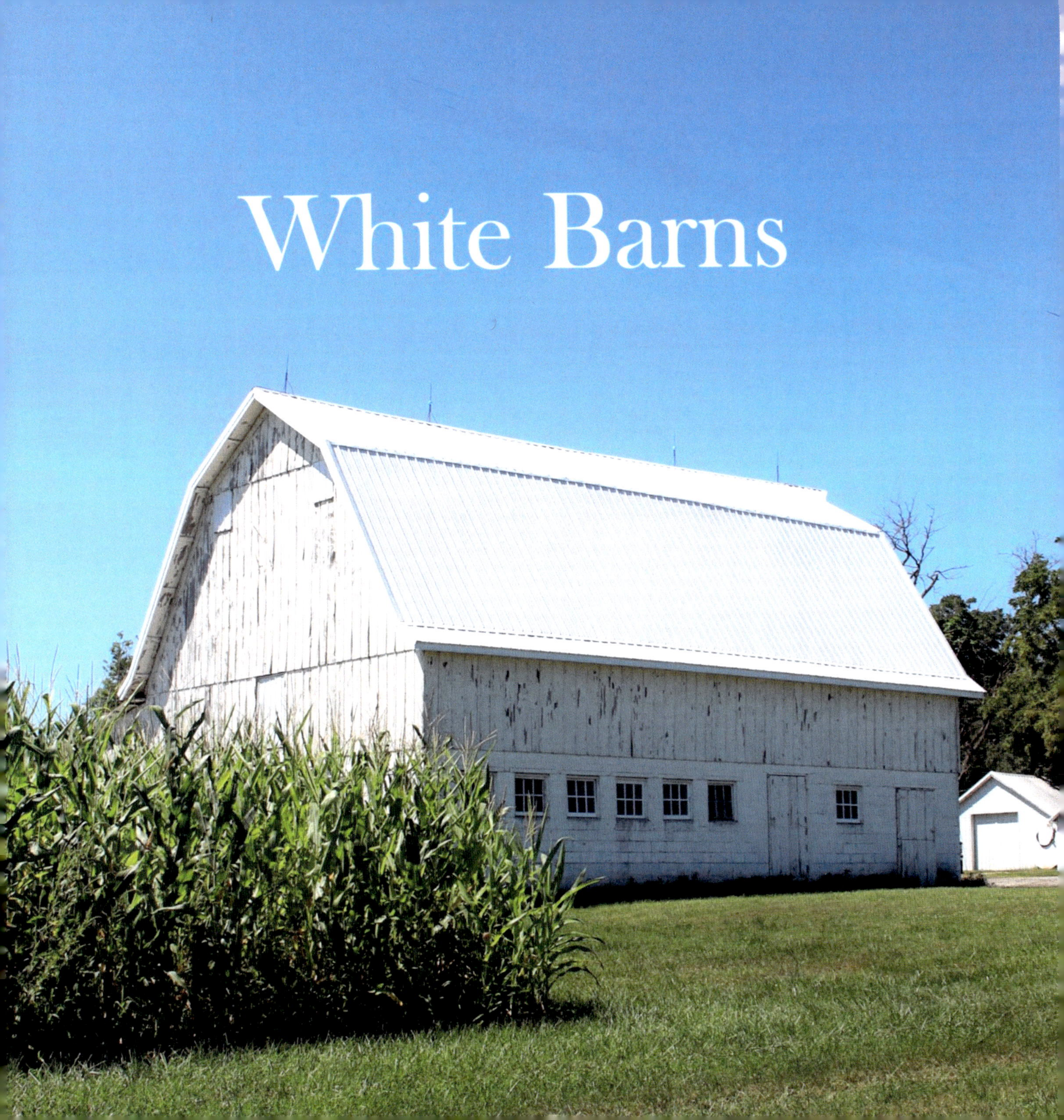

# The National Road

# Spring

INTERNATIONAL

# Lane Farm

# Summer

45

2020
Will County
CANCELLED
HERE
JULY 16 - 19

# The Cobb House

Autumn

NO
OUTLET

# Asides

Winter

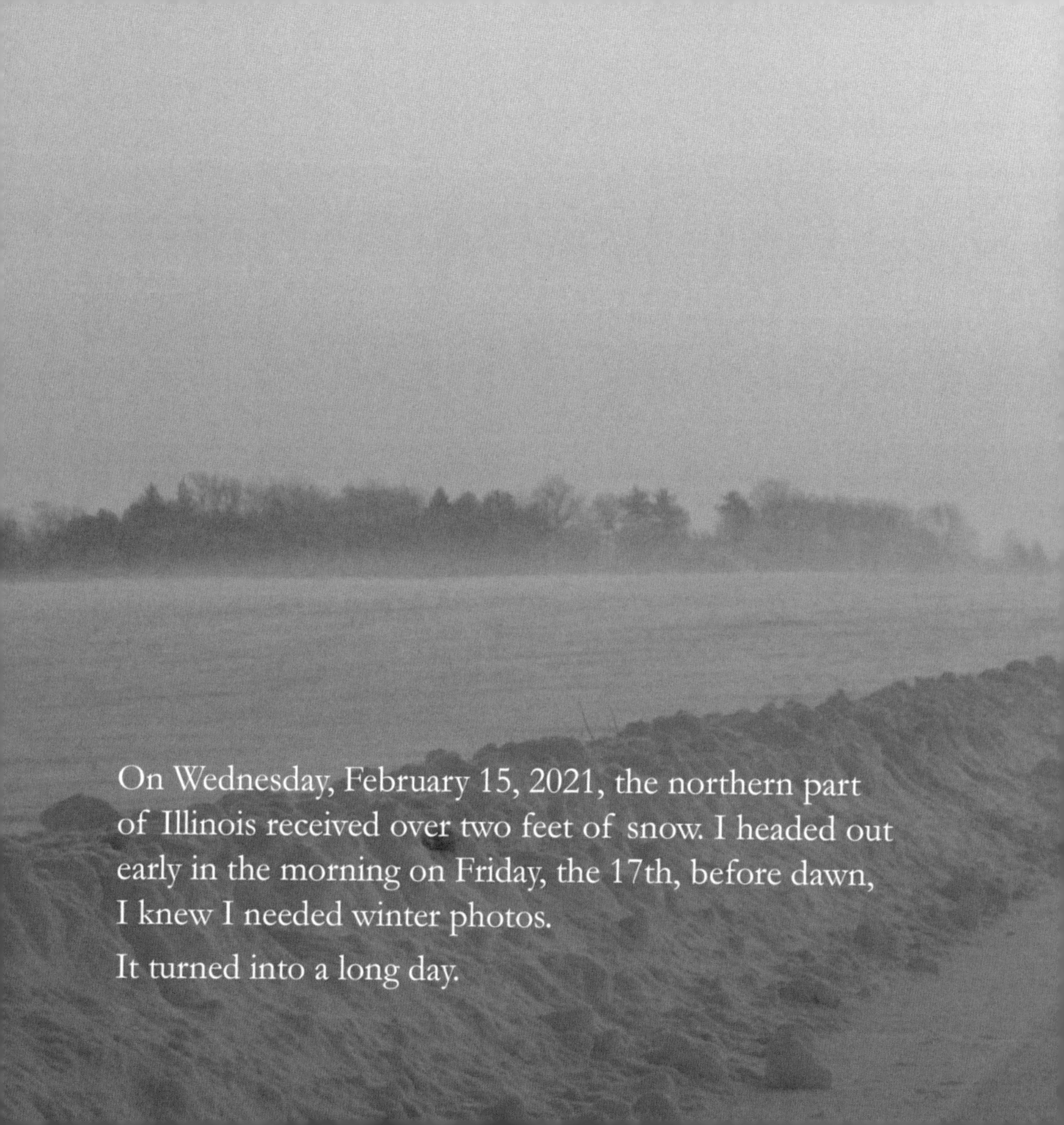

On Wednesday, February 15, 2021, the northern part of Illinois received over two feet of snow. I headed out early in the morning on Friday, the 17th, before dawn, I knew I needed winter photos.

It turned into a long day.

# Final Days

RAILROAD
CROSSING
TRACKS

# Other Books by Kevin McNulty, Sr.

"Around Momence" by Kevin McNulty, Sr.
Copyright © 2007 Kevin McNulty, Sr.
Published by Arcadia Publishing
Charleston SC, Chicago IL, Portsmouth NH, San Francisco, CA
ISBN 978-07358-57289

"Lt. Pat O'Brien" by Kevin McNulty, Sr.
Copyright © 2013 KMC PUBLISHING COMPANY
All rights reserved.
ISBN: 10-0615852114
ISBN-13: 978-0615852119

"The Barns of Kankakee County" by Kevin McNulty, Sr.
Copyright © 2014 KMC PUBLISHING COMPANY
All Photography is the property of Kevin McNulty, Sr.
ISBN-10: 0989796515
ISBN-13: 978-0-9897965-1-4

"Finding Pat O'Brien
Copyright © 2014 KMC PUBLISHING COMPANY
All rights reserved.
ISBN-10: 0989796523
ISBN-13: 978-0-9897965-2

"A Kid from Momence...Growing up After the War"
Copyright © 2015 KMC PUBLISHING COMPANY
All rights reserved.
ISBN-10: 098979654X
ISBN-13: 978-0-9897965-4-5

"Assessing Music Performance...
A Valid System for Measuring Student Performance and Growth"
Copyright © 2017 KMC PUBLISHING COMPANY
All rights reserved.
ISBN 13 978-0-9897965-8-3

"Good Shepherd Manor...A Mission of Compassionate Care"
Copyright © 2021 KMC PUBLISHING COMPANY
All rights reserved.
ISBN: 978-1-732-4091-2-5

## Other Books Published by KMC PUBLISHING COMPANY

"Outwitting the Hun," by Lieutenant Pat O'Brien
Harper W Brothers Publishers, NEW YORK, AND LONDON 1918 - PUBLIC DOMAIN 2013
New York and London, 1918, Public Domain
Republished
Copyright (2) 2013 KMC PUBLISHING COMPANY, All Rights Reserved
ISBN - 0989796507
ISBN - 978-0-9897965-0-7

"Momence...the Making of a River Town," by Kay Hess
Original Copyright © 1985 Kay Hess
Reprinted with permission, The Hess family, Judy (Hess) Hemphill
Copyright © 2015 KMC PUBLISHING COMPANY
All rights reserved.
ISBN-10:0-9897965-6-6
ISBN-13:978-0-9897965-6-9

# ABOUT THE AUTHOR

Kevin McNulty, Sr. published his first book in 2007 at age fifty-seven. Having little experience in publishing, he approached Arcadia Publishing Company, a publisher of neighborhood, local, and regional history of the United States, about doing a book on his hometown of Momence, Illinois. Initially, the publisher turned him down based on the small population of Momence. McNulty told the South Carolina company that he could add two additional towns. He proposed that "Around Momence," include a section on Grant Park, Illinois with a population of 1,500 and an unincorporated area of Kankakee County called St. George which amounted to a Church, a school, and a few homes. Fortunately, Arcadia was unfamiliar with the size of these towns and accepted his proposal. His experience with them taught him how books were put together and guided him in how books are produced.

McNulty's first career out of college was as a high school band and orchestra director. He spent fifteen years teaching before entering the business world and later added five additional years in the classroom between 2011 and 2016 while continuing to write. McNulty credits his life-long interest in history and his opportunity to write at the Chicagoland Chamber of Commerce, as the spark that has kept him writing to this day. He started writing for newsletters, business periodicals and eventually became the publisher of a four-color monthly business magazine called "Chicago Southland," while heading the Chicago Southland Chamber of Commerce. In 2006 he started his own business which took him to China many times and now has publishing as its primary focus.

KMC Publishing Company specializes in writing and producing histories for organizations, institutions, and groups interested in documenting their legacy in book form. McNulty continues to write original works, as well. His most extensive work was an eight-hundred-page historical novel entitled "Lt. Pat O'Brien," © 2013, KMC Publishing Company Inc., about the World War I national hero from Momence who flew for the British before America began using airplanes for war.

A judge for music contests of all types since 1977, McNulty has formulated a unique system of grading for school band, orchestra, and choir directors published in his book, "Assessing Music Performance...A Valid System for Measuring Student Performance and Growth," © 2017, KMC Publishing Company. In 2020, his company created the first online assessment program for music, personnel, or any subjective rating system called "Score-Sheet with Insta-Rank." He is a frequent speaker before business and school groups about subjective assessment.

Kevin McNulty holds a bachelor's degree in Music Education from Illinois State University, a master's degree in Music Education from the University of Illinois and did his MBA studies at Keller Graduate School of Management in Chicago. He resides in Matteson, Illinois with his wife and has three adult children and seven grandchildren.

Made in the USA
Monee, IL
09 June 2021

70815836R10121